Approaches in Highly Parameterized Inversion: GENIE, a General Model-Independent TCP/IP Run Manager

By Christopher T. Muffels, Willem A. Schreüder, John E. Doherty, Marinko Karanovic, Matthew J. Tonkin, Randall J. Hunt, and David E. Welter

Great Lakes Restoration Initiative

Techniques and Methods, Book 7, Section C6

U.S. Department of the Interior
U.S. Geological Survey

U.S. Department of the Interior
KEN SALAZAR, Secretary

U.S. Geological Survey
Marcia K. McNutt, Director

U.S. Geological Survey, Reston, Virginia: 2012

This and other USGS information products are available at http://store.usgs.gov/
U.S. Geological Survey
Box 25286, Denver Federal Center
Denver, CO 80225

To learn about the USGS and its information products visit http://www.usgs.gov/
1-888-ASK-USGS

Contents

Figure

Table

Approaches in Highly Parameterized Inversion: GENIE, a General Model-Independent TCP/IP Run Manager

By Christopher T. Muffels,[1] Willem A. Schreüder,[2] John E. Doherty,[1] Marinko Karanovic,[1] Matthew J.Tonkin,[1] Randall J. Hunt,[4] and David E. Welter[5]

Abstract

GENIE is a model-independent suite of programs that can be used to generally distribute, manage, and execute multiple model runs via the TCP/IP infrastructure. The suite consists of a file distribution interface, a run manager, a run executer, and a routine that can be compiled as part of a program and used to exchange model runs with the run manager. Because communication is via a standard protocol (TCP/IP), any computer connected to the Internet can serve in any of the capacities offered by this suite. Model independence is consistent with the existing template and instruction file protocols of the widely used PEST parameter estimation program. This report describes (1) the problem addressed; (2) the approach used by GENIE to queue, distribute, and retrieve model runs; and (3) user instructions, classes, and functions developed. It also includes (4) an example to illustrate the linking of GENIE with Parallel PEST using the interface routine.

Introduction

A numerical modeling problem that includes multiple independent calculations and requires little effort to separate components of the problem into several parallel tasks is termed "embarrassingly parallel" (Foster, 1995). Parameter estimation, for which finite differences are used to approximate sensitivities, is an example of an embarrassingly parallel problem (Hunt and others, 2010). As is commonly required by parameter estimation, each calibration parameter of interest is perturbed and a corresponding model run executed to assess the observations' sensitivity to the perturbation. Because only one parameter is perturbed in a single run, the results of each model run are completely independent from every other run

needed to calculate the sensitivity of all parameters. Exploiting the embarrassingly parallel nature of parameter estimation software is attractive because individual model run times can be considerable, and simulations of the natural world are often best performed by using a highly parameterized approach (Hunt and others, 2007; Doherty and Hunt, 2010). Therefore, the ability to queue, distribute, and retrieve a set of model runs across a local network or the Internet has the potential to greatly reduce the time of parameter estimation (Hunt and others, 2010). Moreover, the corresponding increase in modeling problem size has created a need for robust and efficient run management for those working on increasingly highly parameterized problems. As such, this work is intended to fulfill a need for projects such as the Great Lakes Restoration Initiative, whereby Great Lake watershed-scale models are calibrated, climate and land-use scenarios are simulated, and uncertainty analyses are performed.

PEST (Doherty, 2010) is a popular parameter estimation program that has recognized this need. It includes a local-network parallel run manager/executer (PPEST/PSLAVE; Doherty, 2010) and is available with a TCP/IP/MPI option provided through BeoPEST (Schreüder, 2009). These programs are PEST-specific; thus, parallelization of model runs can be obtained only in the context for which existing versions of PEST require them. To generalize this capability beyond PEST, a model/program-independent set of tools and functions has been developed that can be used to efficiently manage and execute distributed model runs. This functionality is documented here for a suite of programs called GENIE , which includes

- a run manager, GMAN;

- a run executer, GSLAVE; and

- a routine, GENIE_INTERFACE, that is compiled as part of other programs to handle the exchange of runs with the GMAN run manager.

The GENIE run manager, GMAN, and the GENIE run executer, GSLAVE, are intended to be stand-alone programs that do not require modification by users. Users need only modify their programs to call the provided integration routine that is a conduit/entry point to GMAN, which in turn handles all exchanges regarding the model runs. The run manager

---

[1]S.S. Papadopulos and Associates, Inc.

[2]Principia Mathematica, Inc.

[3]Flinders University and Watermark Numerical Computing.

[4]U.S. Geological Survey.

[5]Computational Water Resource Engineering.

receives runs and distributes them to the different slave computers, where they are executed (fig. 1). The suite is fully compatible with a file distribution interface for automatically distributing model run files across a network.

As discussed by Schreüder and others (2011), functional concerns of a program like GENIE are scalability and load balancing. An efficient run manager should work as well with thousands of slave computers as it does with just a few, with optimization occurring to ensure that all the runs are completed in the least amount of time. Another functional concern is choosing one of the many possible approaches to achieve model independence. GENIE uses the template and instruction file concept as implemented by PEST—the most widely used in parameter estimation codes. The PEST template and instruction file protocols adopted by GENIE facilitate seamless extension of GENIE to any PEST-type application in which ASCII files are used for input and output. Therefore, GENIE can be used to manage and execute model runs for a wide range of purposes including, but not limited to, Monte-Carlo simulations, parametric sweeps, and genetic algorithms. Source code for the interface routine and executables for GMAN and GSLAVE are available for download at *http:// pubs.usgs.gov/tm/tm7c6/*.

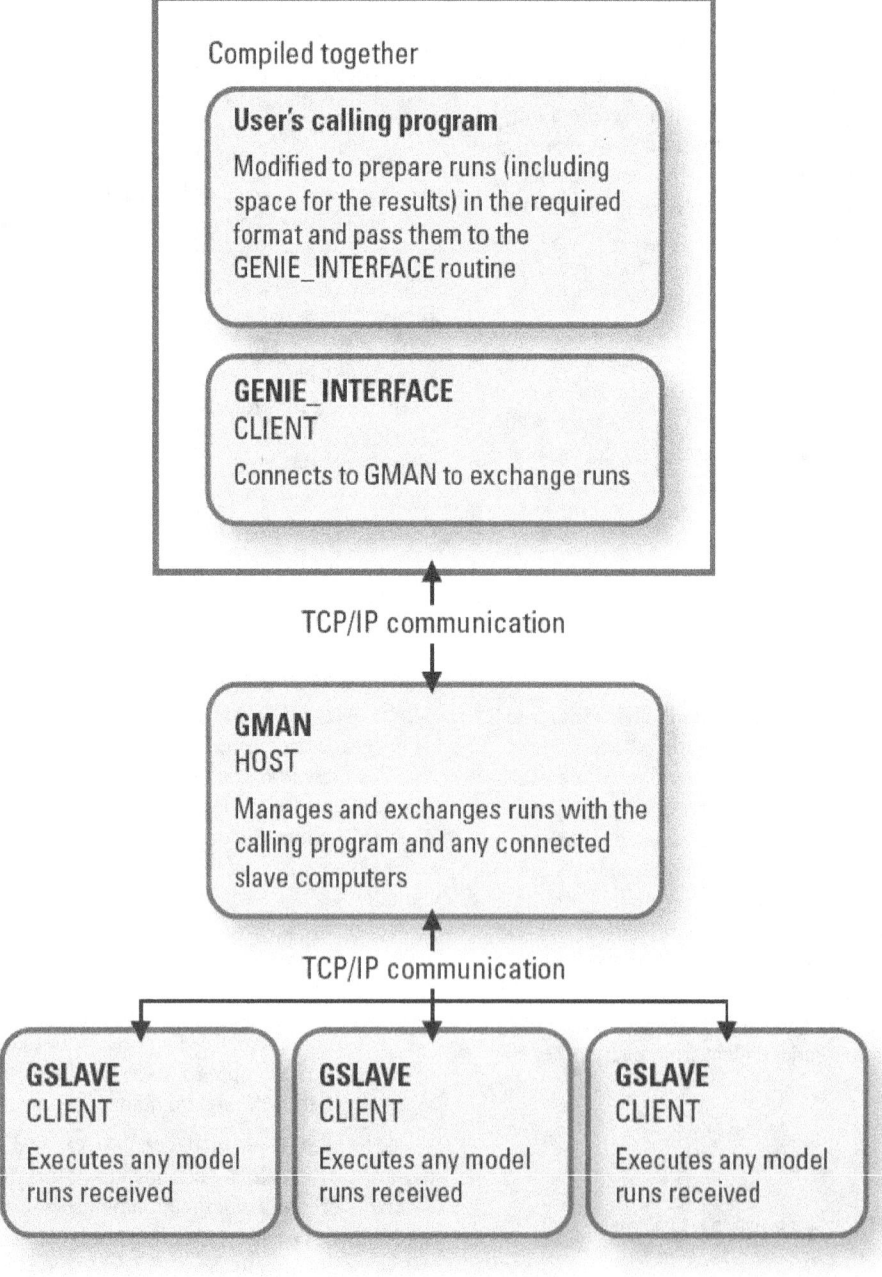

Figure 1.   Flow of communication between the different components of the GENIE suite.

Purpose and Scope

The purpose of this report is to describe GENIE, a suite of programs that can be used to parallelize model runs over a local network or the Internet, using the model independence protocols of PEST (Doherty, 2010). This report is intended for advanced users of programs like PEST. However, each program in the GENIE suite was designed to be simply executed, and beginning users looking to use GENIE with programs for which it is already available (for example, Parallel PEST/ PPEST of Doherty (2010) or PEST++ of Welter and others (2012)) can focus on the input instructions (appendix 1), which forms a quick starting point for the primary GENIE components (GMAN and GSLAVE).

A more sophisticated, higher level purpose of this report is to facilitate the linking of existing software with GENIE to make use of GENIE's powerful programming classes (objects) and tools. The majority of the appendixes detail these classes and tools and provide an example and associated source code, using PPEST, of how to link an existing program with the interface routine. Therefore, although this report provides simple instructions for users (appendix 1), most of its presentation conveys the more advanced concepts of program design in order to facilitate integrating code developed by others. All parameter-estimation-related terminology and concepts use the convention and derivations presented and cited by Doherty (2010) and Doherty and Hunt (2010) and are omitted here for brevity.

Design Concepts

The following descriptions detail the methods employed by the GENIE suite of programs to ensure optimal scalability and to minimize the total execution time for run collections. A "run collection" is the series of runs being managed by the run manager, GMAN. The definition of a "run," the low-level message-passing infrastructure developed for GENIE, and the benefits of object-oriented programming for GENIE are now discussed.

The Structure of a "Run"

The term "run" in this report will be used to describe a single forward model run; for example, a MODFLOW (Harbaugh, 2005) or MT3D (Zheng, 2010) model run. This definition is used because it is consistent with terminology employed by PEST and Monte Carlo simulations. Definition of a run in this manner implies that multiple runs are of the same model type but with differing inputs—as would be the case if doing parameter sensitivity runs such as used to construct the Jacobian matrix. Moreover, the model input values (or parameter values) are what distinguish one run from another. Given this definition, the constituents of a run are listed in table 1.

GENIE itself is not limited to this strict definition of a run. In fact, GENIE was developed to be more flexible, in that a run can be considered any call to a batch or executable file; that is, a set of runs can be structured such that each run is a different model with a unique executable, such as run 1 = a MODFLOW model and run 2 = a MT3D model. The exchange of a run between the calling program and GMAN and GSLAVE version 1.0 is memory intensive. Each constituent is essentially a unique array (see appendix 2) with all runs included in this array. For example, when linking with the interface routine, an NRUN * NPAR element vector is required, where NRUN is the number of runs and NPAR is the number of parameters.

Model Independence

GENIE adopts the template (*.tpl) and instruction (*.ins) file protocols of PEST to realize model independence. Templates of input files are used to interact with models. In the broadest sense, a template file is simply a copy of a model input file with instances of a particular parameter value replaced with the corresponding parameter variable name. Instructions are used to read required result values from model output files. An instruction file is a list of instructions that are interpreted to parse model output files and retrieve the necessary simulated equivalents. Readers are referred to Doherty (2010) for the specifics of template and instruction files, including examples.

Table 1. The constituents of a model run.

Constituent	Description
Executables	Number and name.
Parameters	Number, name, and value. Using the template protocols of PEST, the unique input set for each model run is contained in the parameter values.
Desired model outputs (observations)	Number, name, and value. The instruction protocols of PEST are used to extract model outputs. These are returned to the calling program once a run is complete.
Template and model input files	Number and name. For every template file, the corresponding model input file is required.
Instruction and model output files	Number and name. For every instruction file, the corresponding model output file is required.

C++ and Object-Oriented Program (OOP) Design

GENIE was developed almost exclusively in C++, which is an object-oriented programming (OOP) language and which was used because objects provide a well-established, concise, and extensible means of organizing the functionality of the various programs available to GENIE. Objects are often considered to be overly complicated and sophisticated. Although this can be true, especially for optimal object-oriented programming, their use once coded is not difficult. In addition, in the general sense, the overarching concept is similar to modules in Fortran. Objects are an organizational tool; the code contained within them is still linear—that is, it proceeds from line 1 to the end according to the programmer's logic, just as in Fortran. In addition, because C++ is rooted in C, it is readily compiled with other programming languages, including Fortran, Java, and Python. The example in appendix 3 illustrates how straightforward it is to compile the GENIE_INTERFACE routine with the Fortran program PPEST of Doherty (2010). The ease with which C++ integrates with other languages notwithstanding, it is important to note that language-specific differences may need to be formally addressed. For example, Fortran stores two-dimensional arrays as column-major, whereas C++ stores them as row-major. This difference is highlighted in the example in appendix 3.

Inheritance and polymorphism are perhaps two of the most recognized advantages of OOP. With inheritance, an object—a subclass or derived class—can inherit the attributes and behavior of a pre-existing class (or superclass). Polymorphism allows an object to have more than one form; that is, objects of different types can be defined and invoke methods or properties of the same name, but with type-specific functionality. For example, two computers connected via TCP/IP over the Internet are each nodes in this two-computer network. One computer is the host, whereas the other is the client. Host and client are different types of nodes; the host accepts connections, and the client initiates connection with a host. With C++, a generic node class can be defined with properties and methods common to both the client and host. CLIENT and HOST classes can then be defined to inherit these common attributes, but with behavior unique to their type. From a programming perspective, this polymorphism allows code used to instantiate the host in GMAN to be reused in GSLAVE to instantiate a client. Indeed, at the most basic level, both GMAN and GSLAVE offer the same functionality; what they do with a run is the only difference. Such a design results in an extremely fluid object library that developers can exploit to make robust and sophisticated managers for their own software, more quickly than if they had to program each object independently. Moreover, maintenance, extension, and debugging of the resulting object-orientated code are more efficient than traditional procedural programming.

Message-Passing Infrastructure

GENIE version 1.0 uses only the TCP/IP infrastructure to communicate between its different components. Although other high-level, freely available interfaces exist, the most popular, MPI (MPI Forum, 2009), requires a list of the computers to be used in the network. In GENIE it was more desirable to retain the most flexibility to add and remove nodes (clients) given the dynamic nature of commonly available computing resources, such as cloud computing. Rather than invoking MPI, a low-level message-passing infrastructure was developed to share information between the different components of the program suite. A binary message is made of two parts: a header and a data section. The header is of fixed size and informs the recipient how to interpret the message; specifically, how many bytes are contained in the data portion of the message and the type of that data, whether it is an array of integers or a string or real number. This low-level approach is flexible enough to be used to transmit and process a variety of messages. Developers can use it to create their own messages, furthering the independent nature of GENIE.

The Run Manager—GMAN

The following sections detail the approaches used by GENIE to address the scalability and load-balancing issues discussed by Schreüder and others (2011).

Scalability

The goal of optimizing scalability is to mitigate potential bottlenecks in the job flow. A bottleneck occurs when all the computers in the network wait until one or more nodes work to complete a task. In the case of an embarrassingly parallel problem, this bottleneck is typically the host computer attempting to exchange and process messages with potentially thousands of slave computers (Schreüder and others, 2011). To ensure scalability, the simplest and best solution is to minimize the amount of work the host (or run manager) is responsible for and, following that, to distribute this workload by parallelizing the host itself. In GENIE, GSLAVE utilizes the smart slave concept discussed by Schreüder (2009) to reduce the workload of the run manager. A smart slave is responsible for its own overhead because it writes and reads model files and exchanges only the smallest amount of information needed—exchanging parameter input values and retrieving run results—with the run manager. Utilizing smart slaves, the main tasks left to the run manager are the following:

- exchanging messages with the different slave computers (including disconnects),

- acting upon these messages, and

- distributing/managing the runs (load balancing).

There are three primary approaches to handling multiple client requests:

1. The synchronous approach, in which the host continuously polls each connected socket and processes messages sequentially. It is apparent that this approach can result in a bottleneck because a message must be received and processed in its entirety before another can be handled.

2. The asynchronous approach, in which the host "waits" upon all clients simultaneously. This approach is event driven, so the host no longer polls sockets continuously. Instead, the host is "flagged" by the operating system (OS) when a communication on any socket is pending. However, this approach can result in a bottleneck because the host is still required to process each message sequentially.

3. The multi-threaded approach, in which a unique process or "thread" is spawned to handle each client. Although this approach is well suited to mitigate potential bottlenecks, it is conceivable that a CPU can become hindered by thousands of threads continuously polling their assigned sockets.

GMAN version 1.0 uses a hybrid approach of the asynchronous and multi-threaded approaches above, in which each client is assigned a unique, secondary thread to handle communication asynchronously for the host. With such an approach, the multi-threaded feature allows several messages to be handled simultaneously, while the asynchronous aspect ensures that threads are active only while processing a message. In addition, this approach reduces the responsibility of the host's primary thread to managing and balancing the runs.

Load Balancing

The primary concern of effectively managing the model simulations is optimally balancing the total run load to ensure the total time needed to finish all the runs is minimized. If all the slaves are exactly the same speed and the time to complete each model run is consistent, load balancing is straightforward. However, this cannot be expected for every GENIE application, where it is likely slaves of varying speed will be used. In this case, if faster machines are idling while slower ones are still executing their runs, these same runs will be started on the faster machines. The result is accepted from the first slave to finish and the run terminated on the other. When a potential slave first connects to the run manager, it sends a series of introductory messages including some measure of its speed[1], which GMAN then uses to prioritize or

---

[1] Based on the Linpack Benchmark (Dongarra, 2011) used to rate the top supercomputers. It solves a dense system of linear equations using LU factorization and reports the "MegaFLOPS" rating, or the millions of floating point operations per second.

rank the available slaves. GMAN organizes the complete set of runs into different queues according to their progress state: to do, in progress, complete, or sent. The primary thread in GMAN simply waits for a signal from one of the client threads and then updates the status of each run to reflect its progress.

The Run Executor—GSLAVE

Use of the smart computer slave concept (Schreüder, 2009) in the GENIE suite minimizes the functional role of the run manager, GMAN, making it more scalable. A smart slave is one that is responsible for much of its own run overhead, especially the writing and reading of model files. Information needed for an entire run must be sent by GMAN to GSLAVE; however, only the results and the "true" parameter values need be returned (including the index of the run). Runs are not started as system calls by GSLAVE. Instead they are started as processes. By starting runs in this manner, their operating-system-specific process ID number is readily available and can be used to monitor the progress of that run—and regain control of that run in the event GSLAVE fails (connection with GMAN is lost) and is subsequently restarted. In addition, the process ID provides a mechanism by which orphan processes can be tracked and handled. GSLAVE uses an asynchronous polling approach to monitor communication with GMAN. A single thread is used because GSLAVE is communicating only with GMAN. The polling is timed, which means for a user-specified time GSLAVE is idle and is only "active" for the fraction of a second required to check on the status of the run it is executing. With this approach, GSLAVE is not consuming CPU resources that are better used by a model run.

Limitations of Version 1.0

In order to make GENIE accessible to most applications, the design is extensible. To better extend GENIE, the limitations of version 1.0 are included here.

- Unlike BeoPEST of Schreüder (2009), which includes MPI and TCP/IP communication, GENIE version 1.0 supports only TCP/IP communication.

- Only IP v4 is supported by the GENIE version 1.0 suite. Although relatively trivial to include compatibility to IP v6, there existed no means to test such an extension. Thus, a release of an IP v6 compatible version of GENIE is a topic of ongoing work.

- GENIE version 1.0 is interfaced only with PPEST of Doherty (2010) (also known as GPEST) and PEST++ (Welter and others, 2012).

- The GENIE interface routine is memory intensive: a block of memory is required to hold all of the parameter values and results for each run, though creating a more memory-lean interface is a topic of ongoing work.

- GENIE version 1.0 is programmed only for the Windows®[2] operating system, though porting to UNIX/Linux®[3] is a topic of future work.

- Although this program has been used by the U.S. Geological Survey (USGS), no warranty, expressed or implied, is made by the USGS or the U.S. Government as to the accuracy and functioning of the program and related program material nor shall the fact of distribution constitute any such warranty, and no responsibility is assumed by the USGS in connection therewith.

Summary

The concepts of GENIE, a model-independent suite of programs, are documented herein. GENIE can be used to generally distribute, manage, and execute multiple model runs via the TCP/IP infrastructure. For uses where GENIE has already been combined with the program of interest, the user can simply invoke GENIE by using the instructions in appendix 1. This functionality is available for PPEST of Doherty (2010) and PEST++ (Welter and others, 2012). In addition to the run manager and run executer that are of most use to the majority of users, the full suite also includes a GENIE_INTERFACE routine that can be compiled as part of any program and used to exchange model runs with the run manager. Because communication is via TCP/IP, any computer connected to the Internet can serve in any of the capacities offered by this suite. Model independence is consistent with the existing template and instruction file protocols of the widely used PEST program. Source code for the interface routine and executables for GMAN and GSLAVE are available for download at *http://pubs.usgs.gov/tm/tm7c6/*.

References

Doherty, J., 2010, PEST, Model independent parameter estimation—User Manual (5th ed., with slight additions): Brisbane Australia, Watermark Numerical Computing, 336 p.

Doherty, J.E., and Hunt, R.J., 2010, Approaches to highly parameterized inversion—A guide to using PEST for groundwater-model calibration: U.S. Geological Survey Scientific Investigations Report 2010–5169, 59 p.

Dongarra, J.J., 2011, Performance of various computers using standard linear equations software: Knoxville, Tenn., University of Tennessee, Computer Science Technical Report Number CS-89-85, 107 p., accessed December 13, 2011, at *http://www.netlib.org/benchmark/performance.ps*.

Foster, I., 1995, Designing and building parallel programs: Reading, Mass., Addison-Wesley Pearson Education, 430 p.

Harbaugh, A.W., 2005, MODFLOW-2005, the U.S. Geological Survey modular ground-water model—The ground-water flow process: U.S. Geological Survey Techniques and Methods 6–A16 [variously paged].

Hunt, R.J., Doherty, J., and Tonkin, M.J., 2007, Are models too simple? Arguments for increased parameterization: Ground Water, v. 45, no. 3, p. 254–262.

Hunt, R.J., Luchette, J., Schreuder, W.A., Rumbaugh, J.O., Doherty, J., Tonkin, M.J., and Rumbaugh, D.B., 2010, Using a cloud to replenish parched groundwater modeling efforts: Ground Water, v. 48, no. 3, p. 360–365, doi:10.1111/j.1745-6584.2010.00699.x

MPI Forum, 2009, MPI—A Message-Passing Interface Standard (version 2.2): Accessed December 2009 at *http://www.mpi-forum.org*.

Schreüder, W.A., 2009, Running BeoPEST, *in* Tonkin, M.J., ed., Proceedings, PEST Conference 2009, Potomac, Md., November 1–3, 2009: Bethesda, Md., S.S. Papadopulos and Associates, p. 228–240.

Schreüder, W.A., Muffels, C., Tonkin, M., Doherty, J., Hunt, R.J., and Welter, D., 2011, Efficient use of parallel resources using PEST, *in* MODFLOW and More 2011, Integrated Hydrologic Modeling, International Ground Water Modeling Center, Colorado School of Mines, Golden, Colo., June 6–8, 2011: v. 1, p. 787–791.

Welter, D.E., Doherty, J.E., Hunt, R.J., Muffels, C.T., Tonkin, M.J., and Schreüder, W.A., 2012, Approaches in Highly Parameterized Inversion—PEST++, a Parameter ESTimation code optimized for large models: U.S. Geological Survey Techniques and Methods, book 7, section C5, 47 p.

Zheng, Chunmiao, 2010, MT3DMS v5.3—Supplemental user's guide: Tuscaloosa, Ala., University of Alabama Department of Geological Sciences, Technical Report to the U.S. Army Engineer Research and Development Center, 51 p.

---

[2] "Windows" is a registered trademark of Microsoft Corporation in the United States and other countries.

[3] "Linux" is the registered trademark of Linus Torvalds in the U.S. and other countries.

Appendixes 1 through 3

Appendix 1: Input Instructions

GMAN

GMAN is started by executing the *gman.exe* file. For example, this file can be copied to the desktop (or a shortcut to another folder that contains it) and simply double-clicked to start—actions most easily accessed when running models over a local network only. In the event slave computers are to be used across the Internet, then optional command line switches may be required.

Options

The following are optional switches that can be used to define a specific socket for GMAN to communicate on:

Switch	Description
/ip	Used to specify an IP address to use for communication. Switch is not case sensitive. Only IP v4 is supported currently. Example: /ip 192.168.0.1
/port	Used to specify a PORT to use for communication. This switch is most likely to be required when communicating with slave computers over the Internet. In most cases a specific PORT must be opened in the firewall to allow Internet communication for GMAN. This switch is used to provide that PORT to GMAN. Viable ports are between 1024 and 65535. Switch is not case sensitive. Example: /port 4040

For example, if PORT number 4040 is opened in a firewall, GMAN is started at the command prompt as follows:

```
gman /port 4040
```

GSLAVE

GSLAVE is started by executing the *gslave.exe* file at the command line. The following details the two required switches:

Switch	Description
/host	Used to specify the socket GMAN is listening on. Switch is not case sensitive. Example: /host 192.168.0.1:90210
/name	Used to uniquely identify the slave computer to GMAN. Switch is not case sensitive. Example: /name S-1

The switches can be specified in any order. For example:

```
gslave /interval 1.0 /console on /name S-1 /host 192.168.0.1:4040
```

Options

The following are optional switches that can be used to define a specific socket for GSLAVE to communicate on:

Switch	Description
/ip	Used to specify an IP address to use for communication. Switch is not case sensitive. Only IP v4 is supported in version 1.0. Example: /ip 192.168.0.1
/port	Used to specify a PORT to use for communication. This switch is most likely to be required when communicating with slave computers over the Internet. In most cases a specific PORT must be opened in the firewall to allow Internet communication for GMAN. This switch is used to provide that PORT to GMAN. Viable ports are between 1024 and 65535. Switch is not case sensitive. Example: /port 4040
/interval	The number of seconds GSLAVE is idle before checking on a run. For long model run times, this number can be higher. If this number is unnecessarily low, GSLAVE will consume more CPU resources than it should. For short run times it is important to make this number small (0.1). The default is 10 seconds. Example: /interval 0.1
/console	Indicates whether the RUN should be executed in a visible console window or not. This switch can be either ON or OFF. If it is ON, then the RUN console is visible. The default is OFF. Example: /console ON

Appendix 2: Interacting With GENIE Through GENIE_INTERFACE

The Interface Routine

The interface routine is called GENIE_INTERFACE and is available in a file of the same name with extension *.cpp*. The purpose of this routine is to provide a conduit for information regarding the exchange of model runs and associated results with GMAN for the outside program. The outside program needs to prepare the model run information in the array format required by GMAN and then needs to call the GENIE_INTERFACE routine. The routine returns control to the outside program when all of the runs are complete, after which the outside program can process the results. The routine is written in C++ but contains the necessary external interface for compilation with Fortran. Example source code for such an operation is presented in appendix 3.

The following figure shows the routine declaration, and the table below discusses each required parameter. Parameters are passed by reference (indicated by the * in C++).

```
int GENIE_INTERFACE(int *nrun,
                    int *nexec,
                    char *execnams,
                    int *npar,
                    int *nobs,
                    char *_apar,
                    char *_aobs,
                    double *pval,
                    double *oval,
                    int *ntpl,
                    int *nins,
                    char *_tplfle,
                    char *_infle,
                    char *_insfle,
                    char *_oufle,
                    char *host,
                    char *id,
                    int *ikill)
```

Figure 2–1. The GENIE_INTERFACE routine declaration.

Parameter	Type	Description
NRUN	integer	The number of model runs to be managed and executed.
NEXEC	integer	The number of executable files to be processed as part of a run; currently must be 1 and multiple executables listed in a batch file.
_EXECNAMS	string (NEXEC)	Array, of size NEXEC, listing each executable file.
NPAR	integer	Number of parameters.
NOBS	integer	Number of observations.
_APAR	string (NPAR)	Array, of size NPAR, listing each parameter name.
_AOBS	string (NOBS)	Array, of size NOBS, listing each observation name.

PVAL	double C/C++: (NRUN,NPAR) Fortran: (NPAR,NRUN)	Array, of size NRUN * NPAR, listing each parameter value for each run. Values must be listed in the same order as for _APAR; that is, PVAL(1,1) is the value corresponding to the parameter named _APAR(1,1). Can be 2D array (NRUN,NPAR) or an equivalent 1D array. Because Fortran is column-major, the equivalent Fortran array must be dimensioned and filled as (NPAR,NRUN).
OVAL	double C/C++: (NRUN,NPAR) Fortran: (NPAR,NRUN)	Array, of size NRUN * NOBS, listing each observation value for each run. Values must be listed in the same order as for _AOBS; that is, OVAL(1,1) is the value corresponding to the observation named _AOBS(1,1). Can be 2D array (NRUN,NOBS) or an equivalent 1D array. Because Fortran is column-major, the equivalent Fortran array must be dimensioned and filled as (NOBS,NRUN).
NTPL	integer	Number of template files.
NINS	integer	Number of instruction files.
_TPLFLE	string (NTPL)	Array, of size NTPL, listing the template file names.
_INFLE	string (NTPL)	Array, of size NTPL, listing the model input file names. These file names must be listed in the same order as for _TPLFLE; that is, _INFLE(1) is the model input file whose template is provided by file _TPLFLE(1).
_INSFLE	string (NINS)	Array, of size NINS, listing the instruction file names.
_OUFLE	string (NINS)	Array, of size NINS, listing the model output file names. These file names must be listed in the same order as for _INSFLE; that is, the instructions to read _OUFLE(1) are listed in file _INSFLE(1).
HOST	string	A string containing the socket that GMAN is listening on.
ID	string	A string identifying the calling program to the run manager – for example, "ppest" or "pest++"
IKILL	integer	A flag indicating whether GMAN and all connected GSLAVE instances are to be terminated (0, terminate GMAN; 1, keep GMAN alive). This option is superseded by the GENIE_KILL_GMAN routine.

The interface routine returns a value of non-zero if there is an error and a value of zero otherwise. An example of using the interface with Fortran is provided in appendix 3 of this report.

Terminating GMAN and Any Connected Slave Computers (GSLAVE)

A routine, GENIE_KILL_GMAN, is available with *genie_interface.cpp* that can be called to terminate GMAN and any connected instances of GSLAVE. The necessary external interfaces for compilation with Fortran are provided in the source code. An example of compiling this routine with Fortran is available in appendix 3 of this report. The routine declaration is shown in the figure below, and the table below contains a description of each parameter.

```
GENIE_KILL_GMAN(char *id, char *host)
```

Figure 2–2. The GENIE_KILL_GMAN subroutine declaration.

Parameter	Type	Description
ID	string	A string identifying the calling program to the run manager; for example, "ppest" or "pest++".
HOST	string	A string containing the socket that GMAN is listening on.

Output

The GENIE_INTERFACE routine writes to a log file called *genie_interface.log*. This file is opened in "append" mode, which means log entries are appended to the end of the file. It is up to the developer/user to reset (clear) this file as needed. Each time the GENIE_INTERFACE routine is called, a header "New call for runs --->" is written to the file. The following is a list of log entries recorded to the file:

- Creating synchronization objects

- Preparing executable(s)

- Preparing arrays

- Preparing run collection

- Ready to execute XX runs (where XX is the number of model runs)

- Connecting ID to GENIE RUN MANAGER (where ID is the ID listed in the table above)

- Sending introductory messages to GENIE RUN MANAGER

- Starting RECEIVOR thread to handle communication with GMAN

- Sending RUNS to GMAN

- "Waiting for runs to complete --->" (IDs of completed runs follow)

If there is an error during any of these steps, it is recorded to the log file. Otherwise, upon successful completion of the routine "All runs are complete" is written to the file, followed by a listing of the secondary threads that are terminated. This file can be used to help debug issues when linking with the interface routine.

GENIE Classes

This section details some of the classes developed for the GENIE suite that will be of use to developers interested in linking existing programs with GENIE. The following is a list of the generic classes developed for the GENIE suite of programs. Programming-specific details of these classes are provided in the subsections that follow.

Class (or Object)	Used by	Description
Network programming classes		
NODE	GMAN, GSLAVE, and GENIE_INTERFACE	Contains the bulk of the functionality required to start a node (client or host) using TCP/IP for communication.
HOST	GMAN	The host instance of the NODE class. This class inherits the properties and methods of the NODE class, with unique functionality for opening and listening on a socket to accept incoming client connections.
CLIENT	GMAN, GSLAVE, and GENIE_INTERFACE	The client instance of the NODE class. This class inherits the properties and methods of the NODE class, with unique functionality for connecting to a host.
Message-passing classes		
HEADER	GMAN, GSLAVE, and GENIE_INTERFACE	The header portion of a message.

Class (or Object)	Used by	Description
BUFFER	GMAN, GSLAVE, and GENIE_INTERFACE	A generic buffer class typically used to store the data portion of a message.
MESSAGE	GMAN, GSLAVE, and GENIE_INTERFACE	The low-level message-passing infrastructure developed for GENIE.
Multi-threading classes		
THREAD	GMAN, GSLAVE, and GENIE_INTERFACE	The generic class used to create a new thread. Specific functionality for a thread is contained in other classes.
CONNECTOR	GMAN	An instance of the thread class with specific functionality for accepting connection requests from clients.
RECEIVOR	GMAN, GSLAVE, and GENIE_INTERFACE	An instance of the thread class with specific functionality for receiving messages between two nodes.
SENDOR	GMAN	An instance of the thread class with specific functionality for sending messages between two nodes.
TERMINATOR	GMAN	An instance of the thread class with specific functionality for terminating connections between a client and host and ending the RECEIVOR thread handling communication.
"Run" classes		
MODELRUN	GMAN, GSLAVE, and GENIE_INTERFACE	The definition of a complete model run.
MODELRESULT	GMAN, GSLAVE, and GENIE_INTERFACE	The result portion of a model run, including functionality to send and receive this aspect of a run.
Support classes		
DEFINITIONS	GMAN, GSLAVE, and GENIE_INTERFACE	A generic set of definitions containing specific message types, connection types, and error codes.
COMMANDLINE	GMAN, GSLAVE, and GENIE_INTERFACE	Retrieves the arguments for different command line switches.
EXECUTABLE	GMAN, GSLAVE, and GENIE_INTERFACE	The properties of a model executable; namely, the file name, path, and any arguments.
LINPACK_BENCH	GSLAVE	Returns a measure of the speed of a computer.
SOCKET_UTILITIES	GMAN, GSLAVE, and GENIE_INTERFACE	A set of routines used by the node, client, and host classes to check different aspects of a socket.

The following subsections detail the specific methods and properties of classes that will be of most use to developers; namely, those related to TCP/IP communication, the message-passing infrastructure, and multi-threading. These classes are very general and do not necessarily have to be used together to write a run-management-type program.

Network programming: NODE

The NODE class contains all the common functionality of HOSTs and CLIENTs. The HOST and CLIENT classes inherit these properties and methods and define the functionality for those methods qualified as virtual.

Attribute	Description
terminate	(property – boolean) True if node is to be disconnected. False otherwise.
type	(property – integer) The connection type of the node, where type is defined in definitions h and can be one of the following: TYPE_NOT_SET: Not set. TYPE_MODEL: The calling program; for example, PEST connects to GMAN as type "model" TYPE_SLAVE: A slave-computer /client connection. TYPE_FACE: Option not currently used.
status	(property – integer) The current status of a CLIENT node. Can be one of the following, as defined in definitions h: WAITING_FOR_RUN: Signals the HOST that it is ready to receive a run. RUN_FAILED: Signals the HOST that the run it is charged with completing failed. READY_TO_START: Not used. RUN_COMPLETE: Signals the HOST that the run it is charged with completing was successful and to expect results.
rating	(property – double) The speed rating of the computer as returned by LINPACK_BENCH. This rating is used to sort the collection of CLIENTS according to their relative speed.
name	(property – string) The name of the node.
sock	(property – socket) The socket of the node as set by the operating system for a given IP and PORT.
run_start	(property – collection, list of type clock_t) The start time of each model run executed by the node, where time is the number of seconds since the program was executed.
run_end	(property – collection, list of type clock_t) The end time of each model run executed by the node, where time is the number of seconds since the program was executed.
initialize	(virtual method – integer) The routine called to initialize the properties of a node prior to the node actually starting. Returns 0 if the function fails.
start	(virtual method – integer) The routine that starts the node: In the case of the CLIENT it initiates connection to the HOST, or in the case of a HOST it enters into a listening state to accept incoming connection requests. Returns 0 if the function fails.
stop	(method – void) Disconnects the node and frees any allocated memory.
makenonblocking	(method – integer) Sets the non-blocking property of the socket to true. Returns 0 if the function fails.
settype	(method – void) Sets the type property of the node.

The following methods are used to share the different message types between the different components of GENIE. Developers who require their own messages must first create the message and then use the templates provided by these methods to send the message.

sendrun	(method – integer) Send a run.
sendtype	(method – integer) Send the type of the node.
sendspeed	(method – integer) Send the rating of the node.
sendname	(method – integer) Send the name of the node.
sendcommand	(method – integer) Send a command to a node.
sendstatus	(method – integer) Send the status of the node.

Network programming: HOST

The HOST class inherits the properties and a few of the methods contained in the NODE class. In addition to two unique methods, this class defines the functionality of the virtual methods of the NODE class.

Attribute	Description
initialize	(method – integer) Routine to initialize the properties of the host prior to its actually starting. Returns 0 if the function fails.
start	(method – integer) Routine to start the node. Opens a socket and listens for new connections. Returns 0 if the function fails.
acceptnewconnection	(method – socket) Routine to accept a new client connection. Returns a SOCKET to the client.
waitforcommunication	(method – integer) Routine to a wait a specified amount of time on a set of SOCKETs until communication is pending on at least one of them or the time expires.

Network Programming: CLIENT

The CLIENT class inherits the properties and a few of the methods contained in the NODE class. This class defines the functionality of the virtual methods of the NODE class.

Attribute	Description
initialize	(method – integer) Routine to initialize the properties of a client prior to its actually starting. Returns 0 if the function fails.
start	(method – integer) Routine to start the node. Initiates connection to the HOST. Returns 0 if the function fails.
setsocket	(method – void) Routine to set the socket of the CLIENT.

Message Passing

HEADER

The HEADER class is a set of properties comprising the header of a message. The header is a fixed size (16 bytes currently) that dictates how to read the rest of a message.

Attribute	Description
type	(property – integer) The type of the message. Can be one of the following defined in definitions.h: COMMAND, STATUS_UPDATE, CONNECTION_TYPE, CONNECTION_NAME, CONNECTION_SPEED, or RUN
bytesize	(property – integer) The number of bytes constituting a single element of the buffer portion of a message.
nbytes	(property – integer) The number of elements in the buffer portion of a message.
compression	(property – integer) Currently not used.

BUFFER

The BUFFER class is used to store the data portion of a message.

Attribute	Description
size	(property – integer) The type of the message. Can be one of the following defined in definitions.h: COMMAND, STATUS_UPDATE, CONNECTION_TYPE, CONNECTION_NAME, CONNECTION_SPEED, or RUN
buf	(property – character pointer) The number of bytes constituting a single element of the buffer portion of a message.
alloc	(method – void) Allocate buf to the user-specified size.
dealloc	(method – void) Deallocate buf.
copyfrom	(method – void) Copy source memory block into buf. Uses C memcpy function.
copyto	(method – void) Copy buf to a user-specified memory block. Uses C memcpy function.

MESSAGE

The MESSAGE class is used to prepare, send and receive a message.

Attribute	Description
msgsize	(property – integer) The total number of bytes in the message, including header and data components.
datasize	(property – integer) The number of bytes in the data component of the message.
header	(property – header) The header. See header class.
data	(property – buffer) The data portion of the message. See buffer class.
alloc	(method – void) Wrapper for the buffer class alloc method.
dealloc	(method – void) Wrapper for the buffer class dealloc method.
setmsgsize	(method – void) Routine to set the size of the message. Sets msgsize.
setdatasize	(method – void) Routine to set the size of the data portion of the message. Sets datasize.
waitformessage	(method – integer) Routine to a wait a specified amount of time for communication on a specific CLIENT SOCKET.
receiveheader	(method – integer) Routine to receive the header portion of a message from a CLIENT.
receivedata	(method – integer) Routine to receive the data portion of a message from a CLIENT according to the header.
sendme	(method – integer) Sends a message to a CLIENT.

Multi-threading

The following class works for the Windows OS, the only operating system for which GENIE is available currently.

THREAD

The THREAD class contains most of the code required to start a secondary thread. It can be used to initialize and start a thread—the specific function of the thread is defined in a subclass. By separating the functionality in this manner, it is easy to create threads to handle a variety of purposes or create collections of threads as required for multiple client connections. The GENIE suite implements four different thread types.

Attribute	Description
handle	(property – handle) The handle of the thread (handles are used by Windows to uniquely identify different components).
h_thEnd	(property – handle) A handle to the mutex used to signal termination of the thread.
threaded	(property – unsigned integer) A unique ID assigned to the thread. Not used except in error messages.

execute	(virtual method – integer) The specific routine executed by the thread. What distinguishes the different threads started as part of GENIE is their implementation of this routine; for example, the RECEIVOR and SENDOR threads.
ThreadEntryPoint	(method – static unsigned _stdcall) A required pointer function to the execute routine.
initialize	(method – integer) Initializes the properties.
start	(method – void) Starts the thread.

Events and mutexes are used to communicate between the different threads. An event is simply a flag that can be used to trigger actions in other threads. A mutex is a type of event typically used to sign in and out a shared memory resource to prevent multiple threads from accessing a block simultaneously.

set_event_comm	(method – void) Set event by which this thread can signal the primary thread.
set_mutex_comm	(method – void) Set mutex by which this thread can share memory with the primary thread.
get_parent_end	(method – handle) Returns the primary thread's end event handle. Used to terminate any secondary threads from the primary thread.

Model Run

MODELRUN

The MODELRUN class contains all the properties and methods required to define a model run and exchange it between the different components of the GENIE suite.

Attribute	Description
id	(property, pointer – integer) Assigned by GENIE_INTERFACE, ID is a unique identifier for a run within a run collection.
npar	(property, pointer – integer) The number of parameters.
nobs	(property, pointer – integer) The number of observations.
ntpl	(property, pointer – integer) The number of template files.
nins	(property, pointer – integer) The number of instruction files.
nexec	(property, pointer – integer) The number of executables (currently assumed to be 1).
tplfiles	(property, pointer – string) A pointer to an ntpl-element array that contains the name of each template file.
insfiles	(property, pointer – string) A pointer to an nins-element array that contains the name of each instruction file.
infiles	(property, pointer – string) A pointer to an ntpl-element array that contains the name of each model input file.
outfiles	(property, pointer – string) A pointer to an nins-element array that contains the name of each model output file.

Attribute	Description
parnams	(property, pointer – string) A pointer to an npar-element array that contains the name of each parameter.
obsnams	(property, pointer – string) A pointer to an nobs-element array that contains the name of each observation.
parvals	(property, pointer – double) A pointer to an npar-element array that contains the value of each parameter.
obsvals	(property, pointer – double) A pointer to an nobs-element array that contains the value of each observation.
exec	(property, pointer – EXECUTABLE) A pointer to an instance of the EXECUTABLE class.
init	(method – void) Initializes the pointer variables.
alloc	(method – void) Initializes the pointer arrays.
dealloc	(method – void) Deallocates pointer variables and arrays.
set_exec	(method – integer) Sets the exec properties from a string.
send	(method – integer) Sends the run to the specified socket.
set_frm_msg	(method – integer) Sets the variables and arrays of this class from a message object.
write_input	(method – integer) Only available to GSLAVE. An interface to the PEST Fortran subroutine used to write model input files from template files.
read_output	(method – integer) Only available to GSLAVE. An interface to the PEST Fortran subroutine used to read model output files according to the instructions in the instructions files.
delete_output	(method – integer) Only available to GSLAVE. A system call to delete the model output files from a folder before a run is executed.

MODELRESULT

Similar to the MODELRUN class, but contains only the properties and methods required to send and process model results—the observation values and the adjusted parameter values given the precision they could be written to the model input file.

Appendix 3: Example Application—Interfacing GENIE with PPEST

This section describes in detail the steps taken to link Parallel PEST (PPEST; Doherty, 2010) to GENIE. The link primarily consists of exchanging model runs required and associated results with the run manager, GMAN, using the GENIE_INTERFACE. There are four instances in which PPEST may execute a model run or a collection of runs:

- in the initial model run to get the current objective function,

- during sensitivity matrix calculation,

- during parameter update calculation if using the Levenberg-Marquardt technique, and

- in the final model run using best parameter values.

The PPEST parallel run manager is encapsulated in a subroutine called DORUNS in source file *parpest.f*. This routine serves a role similar to GMAN—it is responsible for distributing and collecting model results as they become available. It is called during each of the four instances listed above. PPEST stores the parameter and observation values for different model runs in direct-access binary files. The runs distributed by the DORUNS routine are constructed from these files. The most significant change to the PPEST program was development of an additional source file called *genie.f* that contains the routines that prepare the PEST data for use with GENIE_INTERFACE and then call this routine. This file is of most interest to developers wishing to link with the GENIE suite.

Most important in the *genie.f* file is a DORUNS-like routine called DORUNS_GENIE. DORUNS_GENIE gathers runs from the various binary files used by PEST to store parameter and observation information, puts them in the necessary arrays required by the GENIE_INTERFACE routine, calls the routine, and finally processes the results in a manner similar to DORUNS so that upon exit from the routine PPEST can continue as normal. All of the modules and subroutines contained within *genie.f* are listed below. The complete source code is listed at the end of this appendix.

Subroutine	Description
GENIE_DATA	Module. Contains the host socket information. These values are set within pest.f from the command line.
checkhost	Routine to verify the host provided is valid. Currently this check is limited to IP:PORT division.
DORUNS_GENIE	Routine equivalent to PEST routine DORUNS for use with GENIE.
END_GENIE	Terminates GMAN and any connected GSLAVES.

All of the changes made to the existing PPEST source code are contained within GENIE preprocessor definitions. The following lists the subroutines that were modified, including the source file they belong to.

Subroutine	File	Description	Change
slavdat1	*parpest.f*	Opens the run manager file and reads the number of slaves.	Initializes the variables set by this routine because the details are not required.
slavedat2	*parpest.f*	Reads part of the run manager file and tests part of the information in it.	Skips details and writes host socket and other details to PPEST run management record (RMR) file.
writslv2	*parpest.f*	Summarizes slave properties to the RMR file.	Added property statements to reflect use of GENIE.
run_pest	*runpest.f*	Main PEST subroutine. Executes the functionality of PEST.	Added calls to GENIE_INTERFACE and GENIE_KILL_GMAN to terminate.
parse_command_line	*pest.f*	Parses the PEST command line.	Modified to get host socket.
pest	*pest.f*	Main program—initialization and call to run_pest.	Modified to indicate use of GENIE.

genie.f Source Code

```
C      -------------------------------------------------------------------
       module GENIE_DATA
       implicit none

       integer               :: port,ikill
       character*25          :: ip
       character*50          :: host

       contains

C      ******************************************************************
       subroutine checkhost(ifail,host_)
       implicit none
C      CTM Mar 2011
C      routine to verify the host provided is valid
C      currently this check is limited to ip:port division
C      ******************************************************************

       integer               :: ifail,idiv,ierr
       character(len=*)      :: host_

       ifail=0

       host=''
       ip=''
       port=-999

C      look for : (separator between port and host)
       idiv=index(host_,':')
       if(idiv.eq.0) then
         ifail=1
         return
       end if

C      parse host into port and ip sections
       ip=host_(1:idiv-1)
       read(host_(idiv+1:len_trim(host_)),*,iostat=ierr) port
       if(ierr.ne.0) then
         ifail=1
         return
       end if

       host=host_

C       write(*,*) trim(ip),',',port,',',trim(host)

       return

       end subroutine checkhost

C      ******************************************************************
       subroutine DORUNS_GENIE(ifail,nrun,pitn,ippp,ippo,ptunit,ptfile,
                              npar,nobs,ntpl,nins,parregfile,obsregfile,
```

```fortran
                               parnam,obsnam,scale,offset,numcom,comlin,
                               tplfle,infle,insfle,outfle,irestart)
         implicit none
c        CTM Apr 2011
c        routine equivalent to PEST routine DORUNS for use with GENIE
c        *****************************************************************

c        Variables associated with call to this routine
         integer                  :: gfail
         integer                  :: ifail,nrun,ippp,ippo,npar,nobs,ntpl,
                                      nins,ptunit,numcom,irestart,pitn
         double precision         :: scale(npar),offset(npar)
         character*20             :: id
         character*200            :: c_parregfile,c_comlin(numcom),ptfile
         character*(*)            :: parregfile,obsregfile,parnam(npar),
                                      obsnam(nobs),comlin(numcom),
                                      tplfle(ntpl),infle(ntpl),
                                      insfle(nins),outfle(nins)

c        Routine specific variables
         integer                  :: jfail,n,ierr,ptcount,titn,itemps,jj,
                                      i,ntodo
         integer,allocatable      :: istatr(:)
         double precision,allocatable :: workvec(:),parval(:,:),obsval(:,:)
         character*11             :: atemp
         character*2000           :: cline

c        GENIE INTERFACE ROUTINE
         integer                  :: GENIE_INTERFACE

         ifail=0

c        null terminate Fortran strings
         id="PPEST"//CHAR(0)
         host=host//CHAR(0)
         ip=ip//CHAR(0)
         c_parregfile=trim(adjustl(parregfile))//CHAR(0)

c        can't simply pull all "words" out of comlin array as
c        each element may contain both an executable name and
c        commandline arguments.
c        add 'fake' end-of-line delimiter to be parsed by C++
         do n=1,numcom
           c_comlin(n)=trim(adjustl(comlin(n)))//'|'
         end do

c        the restart file situation is handled
         allocate(workvec(max(nobs,npar)),istatr(nrun))
         ptcount=0
         istatr=0
         workvec=0.0d0
         if(irestart.eq.1) then
           call ffopen(jfail,-ptunit,'w',ptfile,25,cline)
           if(jfail.ne.0)then
             jfail=0
             irestart=0
```

```fortran
        else
          write(ptunit,iostat=ierr) pitn
#ifdef FLUSHFILE
          call flush(ptunit)
#endif
        end if
      elseif(irestart.eq.2) then
        ptcount=0
        call ffopen(jfail,-ptunit,'r',ptfile,22,cline)
        if(jfail.ne.0) then
          jfail=0
        else
          titn=0
          read(ptunit,iostat=ierr) titn
          if(pitn.eq.titn) then
            do
              read(ptunit,iostat=ierr) itemps
              if(ierr.ne.0) exit
              do jj=1,nobs
                workvec(jj)=-1.1d300
              end do
              istatr(itemps)=-99
              ptcount=ptcount+1
              read(ptunit,iostat=ierr) (workvec(jj),jj=1,nobs)
              if(ierr.ne.0) exit
              call store_parallel_register(jfail,nobs,ippo,itemps,
                                           workvec,obsregfile)
              if(jfail.ne.0) then
                ifail=1
                close(unit=ptunit,iostat=ierr)
                deallocate(workvec)
                deallocate(istatr)
                return
              end if
            end do
            close(unit=ptunit,status='delete',iostat=ierr)
            if(itemps.ne.0) then
              do jj=1,nobs
                if(workvec(jj).lt.-1.0d300)then
                  ptcount=ptcount-1
                  istatr(itemps)=0
                  exit
                end if
              end do
            end if
            call writint(atemp,ptcount)
            write(*,10) trim(atemp)
c#ifdef INTEL
c          icflag=1
c#endif
c          iirun=ptcount
c          ncall=ncall+ptcount
            call ffopen(jfail,-ptunit,'w',ptfile,25,cline)
            if(jfail.ne.0)then
              jfail=0
              irestart=0
```

```
              else
                write(ptunit,iostat=ierr) pitn
                if(ptcount.ne.0)then
                  do i=1,nrun
                    if(istatr(i).eq.-99) then
                      call retrieve_parallel_register
                      (jfail,nobs,ippo,i,workvec,obsregfile)
                      if(jfail.ne.0) then
                        ifail=1
                        close(unit=ptunit,iostat=ierr)
                        deallocate(workvec)
                        deallocate(istatr)
                        return
                      end if
                      write(ptunit,iostat=ierr)i
                      write(ptunit,iostat=ierr)(workvec(jj),jj=1,nobs)
                    end if
                  end do
                end if
#ifdef FLUSHFILE
                call flush(ptunit)
#endif
                if(ptcount.eq.nrun) then
                  deallocate(workvec)
                  deallocate(istatr)
                  return
                end if
              end if
            end if
          end if
        end if
#ifdef LAHEY
10    format('+   Results from ',a,
              ' model runs read from restart file.',/)
#else
10    format('    Results from ',a,
              ' model runs read from restart file.',/)
#endif

c     create run buffer
c     Fortran is column-major order - C++ is row-major
      ntodo=nrun-ptcount
      allocate(parval(npar,ntodo),obsval(nobs,ntodo))
      ntodo=0
      do n=1,nrun
        if(istatr(n).ne.-99) then
          ntodo=ntodo+1
          call retrieve_parallel_register(jfail,npar,ippp,n,
                                  parval(1:npar,ntodo),parregfile)
          do i=1,npar
            if(scale(i).ge.-1.0d35.and.scale(i).ne.0.0d0) then
              if(scale(i).ne.1.0d0.or.offset(i).ne.0.0d0) then
                parval(i,ntodo)=parval(i,ntodo)*scale(i)+offset(i)
              end if
            end if
          end do
```

```
          end if
        end do

c       call GENIE to execute remaining runs
        gfail=GENIE_INTERFACE(ntodo,numcom,c_comlin,npar,nobs,parnam,
                              obsnam,parval,obsval,ntpl,nins,tplfle,
                              infle,insfle,outfle,host,id,ikill)
        if(gfail.ne.1) then
          write(*,20)
          ifail=1
          deallocate(workvec)
          deallocate(istatr)
          deallocate(parval)
          deallocate(obsval)
          return
        end if
20      format(2/,3x,'Genie encountered an error executing runs.')

c       save parameter values in case they have changed due to precision demands
        ntodo=0
        do n=1,nrun
          if(istatr(n).ne.-99) then
            ntodo=ntodo+1
            do i=1,npar
              if(scale(i).ge.-1.0d35.and.scale(i).ne.0.0d0) then
                if(scale(i).ne.1.0d0.or.offset(i).ne.0.0d0) then
                  parval(i,ntodo)=(parval(i,ntodo)-offset(i))/scale(i)
                end if
              end if
            end do
            call store_parallel_register(jfail,npar,ippp,n,
                                         parval(1:npar,ntodo),parregfile)
          end if
        end do

c       store the results and save to restart file
        ntodo=0
        do n=1,nrun
          if(istatr(n).ne.-99) then
            ntodo=ntodo+1
            call store_parallel_register(jfail,nobs,ippo,n,
                                         obsval(1:nobs,ntodo),obsregfile)
            if(irestart.ne.0) then
              write(ptunit,iostat=ierr) n
              write(ptunit,iostat=ierr) obsval(1:nobs,ntodo)
            end if
          end if
        end do
        if(irestart.ne.0) close(unit=ptunit,iostat=ierr)

c       deallocate memory
        deallocate(workvec)
        deallocate(istatr)
        deallocate(parval)
        deallocate(obsval)
```

```
        end subroutine DORUNS_GENIE

c       ************************************************************
        subroutine END_GENIE(ifail)
        implicit none
c CTM May 2011
c terminate GMAN and any connected GSLAVES
c       ************************************************************

c       Routine specific variables
        integer                 :: ifail
        character*20            :: id

c       GENIE INTERFACE ROUTINE
        integer                 :: GENIE_KILL_GMAN

c       null terminate Fortran strings
        id="PPEST"//CHAR(0)
        host=host//CHAR(0)

c       call kill routine
        ifail=GENIE_KILL_GMAN(id,host)
        if(ifail.ne.1) then
          write(*,20)
          ifail=1
          return
        end if
20      format(2/,3x,'Genie encountered an error terminating GMAN.')

        ifail=0

        end subroutine END_GENIE
```

Reference

Doherty, J., 2010, PEST, Model independent parameter estimation—User Manual (5th ed., with slight additions): Brisbane Australia, Watermark Numerical Computing, 336 p.

www.ingramcontent.com/pod-product-compliance
Lightning Source LLC
Chambersburg PA
CBHW081414170526
45166CB00010B/3341